CHEMISTRY Nerd: 1000+ Amazing And Mind-Blowing Facts About Chemistry

© 2023 by Dr. Leo Lexicon

CW01500809

CHEMISTRY NERD: 1000+ Amazing and Mind-Blowing Facts About Chemistry

by

Dr. Leo Lexicon

Also by Dr. Leo Lexicon

AI for Smart Pre-Teens and Teens Ages 10-19: Using Artificial Intelligence to Learn, Think, and Create

AI for Smart Kids Ages 6-9: Discover How Artificial Intelligence is Changing the World

The AI Nerd: Quizmaster Edition Mind-Blowing AI Quizzes that Educate, Entertain and Challenge

Teen Innovators: 30 Teen Trailblazers and their Breakthrough Ideas

Innovation Handbook for Teen Entrepreneurs: Strategies, Tools and Resources to Transform your Vision into Reality

10 Life Hacks Every Teen Should Know: A Comprehensive Guide to Empowerment, Success and Fulfillment in the Teenage Years

Richard Feynman: The Adventures of a Curious Physicist

Nikola Tesla: An Electrifying Genius

John von Neumann: A Giga Brain

Elon: A Modern Renaissance Man

Einstein: The Man, The Myth, The Legend

George Washington: The First American President

Robert Falcon Scott: A Pioneer of Antarctic Exploration

Marco Polo: Intrepid Explorer who Bridged East and West

Captain Cook: The Legendary Seafarer, Navigator, and Explorer

Lewis and Clark: Blazing a Trail to the West

Julius Caesar: The Rise and Fall of Rome's Greatest Leader

Cleopatra: Queen of the Nile

Frida Kahlo: Unbroken Spirit: Artist, Activist, and Icon

QUANTUM COMPUTING for Smart Pre-Teens and Teens Ages 10-19

Quantum Nerd Quizmaster Edition: Quantum Quizzes that Educate, Entertain and Challenge

CHEMISTRY NERD: 1000+ Amazing And Mind-Blowing Facts About Chemistry

PHYSICS NERD: 1000+ Amazing And Mind-Blowing Facts About Physics

BIOLOGY NERD: 1000+ Amazing And Mind-Blowing Facts About Biology

CHEMISTRY NERD:
1000+ Amazing and Mind-Blowing Facts About Chemistry

Welcome to Chemistry Nerd!

Tired of sitting through boring chemistry lectures? This entertaining book is filled with mind-blowing information and a dash of fun to pique your interest in chemistry.

Through an engaging, hands-on examination of chemistry's key concepts, Chemistry Nerd brings the central science to life. Explore the elegance of the Periodic Table, the intricate structures of organic chemistry, and investigate the unique characteristics of carbon that allow it to sustain life.

This book simplifies a wide range of chemical topics—from atomic structure and electrochemistry to acids and bases and spectroscopy—through the use of short but engaging narratives, vivid analogies, and surprising facts.

We are recruiting more nerds who recognize chemistry's aesthetic and transformational value. Please join us!

Dr. Leo Lexicon is an educator and author. He is the founder of Lexicon Labs, a publishing imprint that is focused on creating entertaining and educational books for active minds.

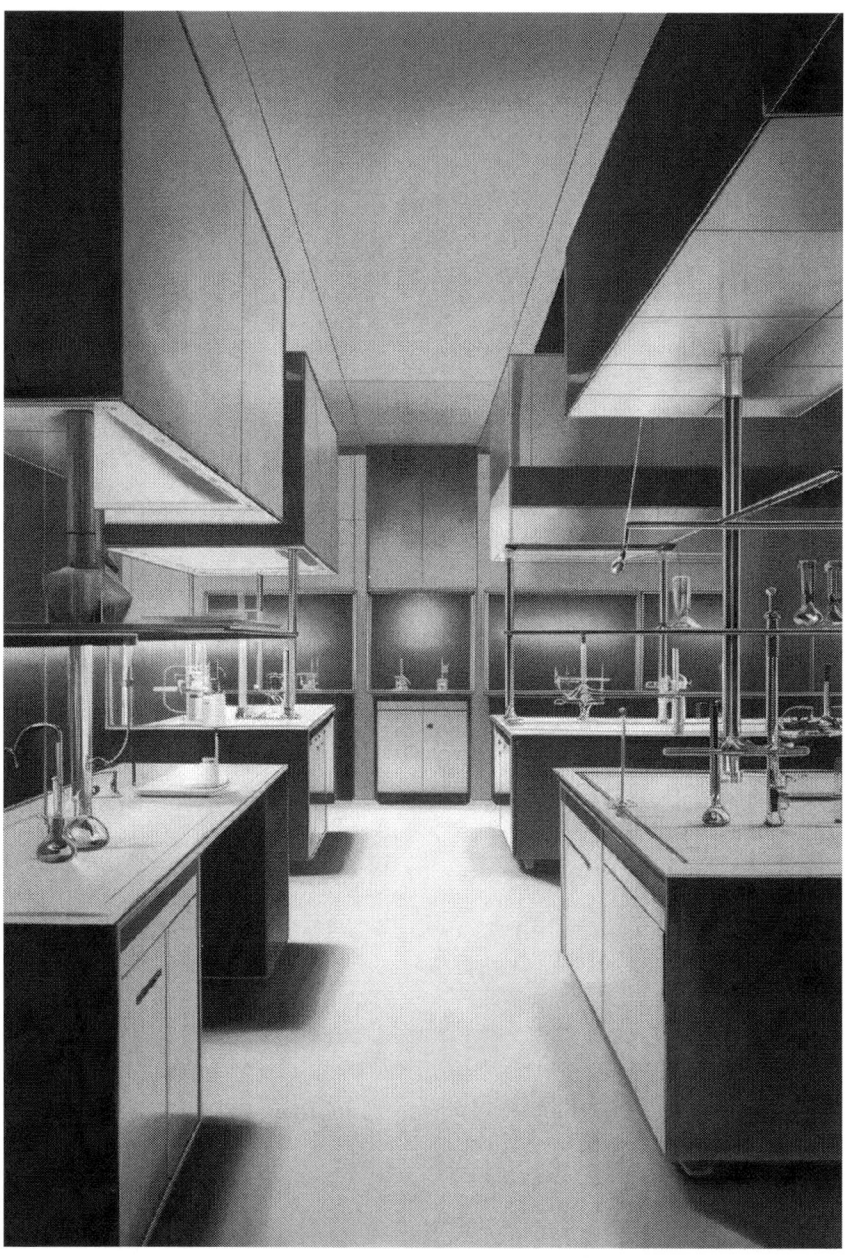

CONTENTS

Chapter 1: The Atomic World

Introduction

Greetings adventurers! Grab your quantum carpets and let's fly through the magical world of atomic science. Our first stop: the interior of the atom!

Protons, Neutrons and Electrons

All matter in the universe is made up of tiny particles called atoms. Atoms are made up of three even smaller particles: protons, neutrons, and electrons. Protons and neutrons are located in the center of the atom, in a region called the nucleus. Electrons orbit around the nucleus in a cloud.

The number of protons in an atom is called its atomic number. The atomic number uniquely identifies each element. For example, all hydrogen atoms have one proton, all helium atoms have two protons, and all carbon atoms have six protons.

The number of protons plus the number of neutrons in an atom is called its mass number. Atoms of the same element can have different numbers of neutrons, and these are called isotopes. For example, the most common isotope of hydrogen has one proton and no neutrons, but there is also an isotope of hydrogen with one proton and one neutron.

The electrons in an atom are arranged in energy levels. The energy levels are filled from the lowest level to the highest level, and the Aufbau principle states that electrons will fill the lowest energy level available before filling a higher energy level.

The Periodic Table

The periodic table is a way of organizing the elements based on their atomic number and chemical properties. The elements are arranged in rows (periods) and columns (groups). Elements in the same group have similar chemical properties. For example, all of the alkali metals in group 1 are very reactive and have one valence electron.

Fig. The Periodic Table

The periodic table also shows trends in atomic properties. For example, as you move from left to right across a period, the atoms become smaller and more electronegative. Electronegativity is a measure of how strongly an atom attracts electrons.

Now let's look at some of the intriguing but ever-elegant aspects of the periodic table is organized:

- The periodic table arranges the elements by increasing atomic number and groups them by chemical properties.
- In 1869 Dmitri Mendeleev left gaps in his early periodic table predicting unknown elements like gallium and germanium. What vision!
- Henry Moseley later showed atomic number was the defining property of elements, establishing the basis of the modern periodic table.
- The table has 7 horizontal periods or rows reflecting the filling of electron shells from 1 to 7. New periods start when new electron shells begin.
- There are 18 vertical groups or families with elements displaying similar chemical properties and valence electron configurations.
- The main groups are numbered 1 through 18. Transition and inner transition metals form in between groups.
- Group 1 alkali metals like lithium and sodium are extremely reactive metals with a single valence electron.
- Group 2 alkaline earth metals such as calcium and barium are also reactive metals with two outer electrons.
- Group 17 halogens including chlorine and iodine are nonmetals one electron short of a stable octet.
- Group 18 noble gases like neon and argon have full valence shells and low reactivity.
- Period 1 has only 2 elements, hydrogen and helium, because the first shell fills so quickly.
- Atomic size decreases moving left to right across a period. Adding protons tightens the nucleus as electrons fill inner shells.
- Density and melting point also show trends across periods and down groups. Fun patterns emerge!
- Metals cluster on the left of the periodic table, nonmetals on the right. Metalloids like silicon have intermediate properties.
- Electron affinity describes the energy change when an atom gains an electron. Chlorine welcomes an extra electron with open arms!

- Electronegativity measures an atom's attraction for electrons in a bond. Fluorine wins most electronegative, francium least.
- Ionization energy is the energy needed to remove an electron from an atom. Noble gases cling tightly to their stable octets.
- Ionic radius depends on the outer electrons. Cations are smaller than their parent atom, anions larger.

Let's consider some real elements to see these atomic principles in action!

- Hydrogen has atomic number 1. Its single electron occupies the 1s orbital. Simple!
- Helium has two electrons that fill the first shell. Its full outer shell makes it stable and inert.
- Lithium, atomic number 3, has two inner electrons and one outer electron. It's the first alkali metal.
- Carbon has four outer electrons that can form four covalent bonds, making complex organic molecules possible!
- Neon, with 10 protons and 10 electrons arranged in filled shells, is a happy inert gas. It glows red-orange in signs.
- Iron has 26 electrons distributed across multiple inner shells and orbitals. This complex configuration leads to the many oxidation states of iron.
- Copper's outer d shell allows it to easily lose an electron to form Cu+ ions, explaining copper's role in electrochemistry and conductivity.
- Krypton's 36 electrons fill the outer shell making it another noble gas. Krypton lasers emit a bright blue-green light.

Amazing Facts from the Atomic World

- Atoms are incredibly tiny, around 0.1 to 0.5 nanometers. You could line up 10 million atoms across the width of a human hair!

- Atoms consist of a dense central nucleus surrounded by a probabilistic electron cloud. It's like a star encircled by rings of satellites.
- The nucleus contains positively charged protons and neutral neutrons, while negative electrons occupy energy shells and orbitals. Opposites attract!
- Protons and neutrons have around 2000 times the mass of electrons. The nucleus holds over 99.9% of an atom's mass.
- Protons and neutrons are made of even smaller particles called quarks! Protons contain two up quarks and one down quark.
- The strong nuclear force overcomes electric repulsion between protons, gluing the nucleus together. Without it, atoms would fly apart!
- Atomic mass is the total number of protons and neutrons. Carbon-12 has 6 protons and 6 neutrons for a mass number of 12.
- Isotopes are versions of elements with the same atomic number but different numbers of neutrons.
- Radioactive carbon-14 is an important isotope used in radiocarbon dating of ancient artifacts.
- Uranium has no stable isotopes, only radioactive ones! Uranium-235 is key for nuclear power.
- Atomic number is the number of protons, defining an element's identity. 6 protons means you're carbon!
- Electrons have dual wave-particle duality described by quantum mechanics. They exist in probabilistic orbitals surrounding the nucleus.
- Electrons behave like waves in the famous double slit experiment, producing interference patterns. Amazing!
- Electrons fill orbitals following the Aufbau principle and Pauli exclusion principle. Orderly filling maximizes electrons while minimizing energy.
- The principal quantum number n labels electron shells. Shell 1 holds 2 electrons, shell 2 has 8. Shells expand to hold 18, 32, and more electrons.
- The closest shell to the nucleus is called the 1s orbital and can only hold 2 electrons with opposite spin.

- The four quantum numbers - n, l, ml, and ms - describe the electron configuration. l is the orbital shape, ml its orientation in space, ms the spin.
- No two electrons in an atom can have the exact same four quantum numbers according to the Pauli exclusion principle. Electrons are territorial!
- The maximum number of electrons in any shell is $2n^2$ where n is the principal quantum number. This restricts the number of elements.
- Core electrons in inner shells are tightly bound to the nucleus. Valence electrons in the outer shell determine chemical reactivity.
- Valence electrons involved in bonding are called bonding electrons. Non-bonding valence electrons are called lone pairs.
- Elements with filled or empty outer electron shells are super stable, like inert gases. Elements seeking full outer shells are reactive.
- Transition metals have partially filled inner d orbitals that allow many oxidation states for complex chemistry.
- Electron transitions between orbitals release photons of light at specific energies, producing unique atomic emission spectra.
- Each element's atomic spectrum is like a unique fingerprint used to identify it. Hydrogen's fingerprint has red, blue, and violet lines.
- Emission spectra were key evidence for the quantum model of the atom with electrons occupying discrete energy levels.

Recent Breakthroughs

- Scientists have developed new methods for creating and manipulating atoms and molecules at the single-atom and single-molecule level. This could lead to new ways to develop new materials, drugs, and other technologies.
- Researchers are using atomic chemistry to study the behavior of atoms and molecules in extreme environments,

such as high temperatures and pressures. This research could help us to better understand the chemistry of planets and stars, as well as the chemistry of materials that are used in extreme applications.

- Scientists are using atomic chemistry to study the chemistry of life. For example, researchers are using atomic chemistry to understand how enzymes work and how proteins fold into their three-dimensional structures.
- Atomic chemistry is being used to develop new ways to diagnose and treat diseases. For example, scientists are using atomic chemistry to develop new drugs that can target and destroy cancer cells.

We've covered a lot of ground exploring the basics of atomic science! From quirky quantized electrons to the elegant periodic table, atoms are full of wonders. Let's delve deeper into electron configurations, nuclear properties, and the quantum atom in the chapters ahead!

Chapter 2: Chemical Bonding and Reactions

Introduction

Welcome back chemistry explorers! Let's continue our adventure into the world of chemical bonding and reactions.

Types of Bonds

Chemical bonds are the forces that hold atoms together to form molecules and compounds. There are three main types of chemical bonds: ionic, covalent, and metallic.

Ionic bonds form between atoms with different electronegativities. Electronegativity is a measure of how strongly an atom attracts electrons. Atoms with high electronegativities tend to gain electrons, while atoms with low electronegativities tend to lose electrons. When an atom gains or loses electrons, it becomes an ion. Ions with opposite charges attract each other, forming ionic bonds.

Covalent bonds form between atoms that share electrons. This type of bond is most common between nonmetals. When two atoms share electrons, they both become more stable.

Metallic bonds form between metal atoms. Metal atoms have a relatively large number of valence electrons (electrons in the outermost energy level). These valence electrons are loosely held and can easily move from atom to atom. This sea of mobile electrons is what holds metal atoms together in metallic bonds.

In addition to chemical bonds, there are also intermolecular forces. Intermolecular forces are the attractive forces between molecules. The three main types of intermolecular forces are dipole-dipole interactions, hydrogen bonding, and London dispersion forces.

Lewis Structures and VSEPR Theory

Lewis structures are a way of representing the valence electrons in an atom or molecule. Valence electrons are the electrons in the outermost energy level of an atom. Lewis structures can be used to predict the shape of molecules and the types of bonds that will form between atoms.

VSEPR theory (valence shell electron pair repulsion theory) is a theory that explains the three-dimensional shape of molecules. It looks at how the electrons around the central atom of a molecule arrange themselves to minimize repulsion between the electrons.

For example, let's look at methane, $CH4$. The carbon atom in the center has four bonds to the hydrogen atoms. But carbon also has two lone pairs of electrons that do not form bonds.

According to VSEPR, these six electron groups (the four bonding pairs and two lone pairs) will try to get as far away from each other as possible to minimize repulsion. This results in the two lone pairs moving to opposite ends of the central carbon atom, while the four hydrogens spread out evenly in a tetrahedron shape.

So methane takes on a tetrahedral molecular geometry thanks to VSEPR theory accounting for how the six electron pairs arrange themselves around the central carbon to stay as far apart as possible. We can predict other molecular shapes by applying VSEPR theory and considering the number and type of electron pairs.

Chemical Reactions

Balancing chemical equations is important because it allows us to predict the amount of reactants and products in a chemical reaction. A balanced chemical equation has the same number of atoms of each element on both the reactants and products sides of the equation.

Chemical reactions can be classified into different types based on the changes that occur during the reaction. The four main types of chemical reactions are synthesis, decomposition, single replacement, and double replacement reactions.

Chemical bonds form when atoms share or transfer electrons. Bonding allows atoms to satisfy the octet rule and attain stable electron configurations. There are several main types of bonds:
- Ionic bonds form between metal and nonmetal atoms, where electrons are completely transferred to form ions.

For example, sodium (Na) has 1 valence electron it can donate. Chlorine (Cl) has 7 valence electrons and wants 1 more. Sodium donates its electron to chlorine, forming $Na+$ and $Cl-$ ions. The opposite charges attract, creating an ionic bond.
Ionic compounds like sodium chloride (NaCl) usually form crystal lattice structures, with alternating positive and negative ions. Ionic bonds result from electrostatic attraction.
- Covalent bonds involve shared electron pairs between nonmetal atoms.

In hydrogen (H2), each hydrogen shares its single electron with the other, forming a covalent bond. This allows both to satisfy the octet rule and attain a stable electron configuration.

In water (H2O), the oxygen and hydrogens share electrons to fill the outer shell of oxygen and each hydrogen. The covalent bonds result from overlap of atomic orbitals.

- Metallic bonds form between metal atoms where electrons float freely amongst positive metal ions.

Sodium metal has one valence electron that leaves the atom. The remaining electrons are delocalized, creating a "sea of electrons" moving freely. This accounts for metals conductive and malleable properties.

- van der Waals forces are weak intermolecular attractions between molecules or atoms.

Temporary fluctuations in the electron clouds of atoms can induce slight polarities that attract nearby atoms or molecules. van der Waals forces increase with larger atoms or molecules.

These minor forces explain phenomena like noble gas condensation into liquids, even though noble gases do not form covalent or ionic bonds. Tiny van der Waals add up!

- Hydrogen bonds are a powerful intermolecular force between molecules with H bonded to N, O or F.

In water, the H forms a partial positive charge, while O has a partial negative charge. This creates hydrogen bonding between water molecules, which accounts for many of water's properties.

DNA base pairs are held together by hydrogen bonds between the nitrogenous bases. This hydrogen bonding stabilizes the double helix structure.

To recap:
Lewis structures illustrate valence electron arrangements in molecules:

- Each dot represents one electron in the valence shell.
- Shared electron pairs represent covalent bonds between atoms.
- For example, H2O has 2 dots around oxygen, representing its 6 total valence electrons. The oxygen shares two

electron pairs with the hydrogens, forming two single covalent O-H bonds.
- Double bonds share two electron pairs and triple bonds share three.
- Oxygen gas, O_2, has a double bond with four shared electrons between the oxygens.

Molecular geometry or shape is predicted by VSEPR theory:
- Electron pairs and lone pairs repel each other, assuming positions that minimize repulsion.
- In CO_2, the two C=O double bonds are 180 degrees apart to minimize repulsion, giving CO_2 a linear shape.
- In NH_3, the three N-H bonds are approximately 120 degrees apart, with the lone pair on nitrogen pushing the hydrogens down to form a trigonal pyramidal shape.

Chemical reactions rearrange the bonding between atoms to form new substances:
- Chemical equations represent reactions with reactants on the left and products on the right.
- Coefficients balance the number of atoms before and after the reaction.
- For example, combustion of methane:
  $CH_4 + 2O_2 -> CO_2 + 2H_2O$
- Balanced equations follow the law of conservation of matter. Atoms aren't created or destroyed, just rearranged!
- Some reactions like $H_2 + Cl_2 -> 2HCl$ are reversible, with reactants reforming from the products.
- Precipitation reactions occur when cations and anions in solution combine to form an insoluble solid precipitate, like Ag^+ and Cl^- forming solid $AgCl$.
- Acid-base reactions involve transfer of H^+ protons. $HCl + NaOH -> H_2O + NaCl$
- Oxidation is the loss of electrons, reduction is the gain. Redox reactions transfer electrons between reactants.

Chemical reactions can be further categorized:
- Synthesis reactions combine simpler reactants to form more complex products. For example:

N2 + 3H2 -> 2NH3 (nitrogen reacts with hydrogen to form ammonia)
- Decomposition reactions break complex reactants into simpler products:
  2H2O -> 2H2 + O2 (water decomposes into hydrogen and oxygen)
- Single replacement reactions swap one atom for another:
  Zn + 2HCl -> ZnCl2 + H2 (zinc replaces hydrogen in HCl)
- Double replacement reactions exchange ions between ionic compounds:
  NaOH + HCl -> NaCl + H2O (sodium and chloride switch partners)
- Combustion reactions involve burning fuels in oxygen:
  CH4 + 2O2 -> CO2 + 2H2O

Understanding the different types of chemical bonds and reactions is essential for exploring all of chemistry! Let's continue marveling at the microscopic world dictating macroscopic phenomena. More bonding and reaction adventures await! Let us close with these five quotes that get to the crux of the matter:

- "Atoms are not things, they are only the pattern of behaviors." - Richard Feynman
- "In chemical reactions, electrons play a game of musical chairs." - John Emsley
- "Bonding is nature's way of keeping molecules together when logic says they should fall apart." - K.C. Nicolaou
- "Chemical bonds want to be shared. Atoms want full outer shells. This explains reactivity." - John Gillespie
- "Chemical reactions write the storybook of our world, from metabolism to metallurgy." - Sam Kean

Recent Breakthroughs

- Scientists have developed new theories and computational methods to predict and understand chemical bonding and reactions more accurately. This could lead to new ways to design new materials, drugs, and other chemicals.

- Researchers are using new techniques to study chemical bonding and reactions in real time. This research could help us to better understand how chemical reactions happen and how to control them.
- Chemists are developing new ways to catalyze chemical reactions. Catalysts are substances that speed up chemical reactions without being consumed themselves. New catalysts could lead to more efficient and sustainable chemical processes.
- Scientists are developing new catalysts that can be used to split water into hydrogen and oxygen, which can then be used to produce fuel.

Chemical bonds and reactions are essential for understanding chemistry. By understanding how atoms interact with each other to form bonds and compounds, we can explain and predict the behavior of matter. With dedication and practice, you can become an expert at understanding these bonds and reactions.

Chapter 3: Acids and Bases

Introduction

Grab your goggles as we cautiously enter the world of acids and bases. Things are about to get messy!

Acids and bases are two important classes of chemicals that play a vital role in many chemical and biological processes.
- Acids are substances that donate protons (H+) in water.
- Bases are substances that accept protons (H+) in water.

The pH Scale

The pH scale measures how acidic or basic a solution is. It ranges from 0 to 14, with 7 being neutral. pH values below 7 are acidic, while pH values above 7 are basic. pH is calculated as the negative log of the hydrogen ion (H+) concentration. A low pH means high acidity and high [H+].

For example, lemon juice has a pH of around 2 due to its citric acid content. Pure water has a pH of 7. Ammonia has a pH of around 11 due to its ability to gain H+ ions. The lower the pH, the more acidic and higher the [H+].

Acid and base strength depends on how easily an acid donates H+ ions or a base accepts H+ ions. Strong acids like hydrochloric acid (HCl) or nitric acid (HNO3) completely dissociate into H+ and anions in water.

Weak acids like acetic acid (HC2H3O2) only partially ionize. Strong bases fully dissociate into hydroxyl ions (OH-), while weak bases partially dissociate.

For example, HCl is a strong acid that ionizes completely:
HCl + H2O → H3O+ + Cl-

Acetic acid is a weak acid that only partially ionizes:
HC2H3O2 + H2O ⇌ H3O+ + C2H3O2-

As we can see, the equilibrium favors the reactants.

Acid-base reactions involve the reaction between an acid and a base to form water and a salt. For example, mixing hydrochloric acid and sodium hydroxide results in an acid-base reaction:
HCl + NaOH → H2O + NaCl

The H+ from the acid reacts with the OH- from the base to form water. The other product is table salt, NaCl.

Titration analysis precisely determines acid or base concentration. A known concentration of one solution is slowly added to a solution of unknown concentration until neutralization is reached. Indicators signal the endpoint. The volumes used allow calculation of the unknown concentration. Titrations require accurate volume measurements and careful handling of reagents.

Applications of Acids and Bases

Acids and bases are used in a wide variety of applications, including:
Industry: Acids and bases are used in many industrial processes, such as the production of fertilizers, plastics, and pharmaceuticals.
Medicine: Acids and bases are used in many medical applications, such as the treatment of indigestion, ulcers, and cancer.
Household products: Acids and bases are used in many household products, such as cleaning products, foods, and beverages.

Acids and bases are two important classes of chemicals that play a vital role in many chemical and biological processes. They have a wide variety of applications in industry, medicine, and everyday life.

Amazing Facts and Quotes

- "An acid is a substance that donates protons, a base is a proton acceptor." - Gilbert Lewis
- Acids release H+ protons in water. The hydrogen ion is just a bare proton since hydrogen has only one electron.
- Bases accept H+ protons in water. Bases release hydroxide (OH-) ions which combine with H+ to form water.
- Arrhenius originally defined acids as compounds that release H+ in water, and bases as compounds that release OH-.
- Common acid examples: hydrochloric acid (HCl), citric acid (in lemons), sulfuric acid (battery acid).
- Common bases: sodium hydroxide (NaOH), calcium hydroxide (Ca(OH)2), ammonia (NH3).
- "The pH scale measures how acidic or basic a solution is." - Michael Munowitz
- pH ranges from 0 (most acidic) to 14 (most basic), with 7 being neutral.
- pH = -log[H+]. The more acidic, the higher the [H+], the lower the pH.

- pH of battery acid ~ 0, lemon juice ~ 2, neutral water ~ 7, household bleach ~ 12.
- Litmus paper indicates pH by changing color. Red = acidic, blue = basic.
- Universal indicator and pH probes precisely measure pH. Color or voltage changes indicate pH.
- "Strong acids fully dissociate in water, weak acids only partially dissociate." - Elizabeth Cussler
- Strong acids like HCl or HNO3 ionize completely, giving lots of free H+.
- Weak acids like acetic acid (HC2H3O2) only partially ionize and donate some H+.
- Strong bases like NaOH or KOH fully dissociate into ions in water. Weak bases partially dissociate.
- The strengths of acids and bases depends on the degree of ionization in water.
- "Acids and bases drive many reactions through proton transfer." - Christopher Hoang
- In neutralization reactions, acids and bases react to form water and a salt. HCl + NaOH -> H2O + NaCl
- Titrations measure acid or base concentration by carefully adding one solution to another until neutralization occurs.
- Indicators like phenolphthalein signal the endpoint where all acid or base has been consumed by color change.
- Buffers resist pH change by absorbing excess H+ or OH-. Blood acts as a buffer to maintain our normal pH.
- Acid rain forms when sulfur dioxide and nitrogen oxides dissolve in water to make weak acids, lowering the rain's pH. Not fun for forests!
- Antacids like Tums contain bases to neutralize stomach acid. Bases taste bitter, while acids taste sour.
- "Acids and bases are crucial for organic reactions, industry, and biology." - Janice Gorzynski Smith
- Soap is made from reaction of fatty acids with sodium hydroxide. Acidic ingredients help give food flavor.
- Acids dissolve certain minerals to extract metals, make fertilizers, or produce pigments and dyes.

- Digestive acids facilitate breakdown of food. Nucleic acids like DNA store genetic information.

Let's recap:

- Acids release protons, bases accept protons.
- pH measures acidity or basicity on a scale from 0 to 14.
- Strong acids and bases ionize completely in water, weak acids and bases partially ionize.
- Acids and bases neutralize each other through proton transfer reactions.
- Buffers resist pH change. Acids and bases are essential for life and industry.

Recent Breakthroughs

- Scientists have developed new types of acids and bases that are more efficient and sustainable than traditional acids and bases. These new acids and bases could be used in a variety of applications, such as catalysis, energy storage, and environmental remediation.
- Researchers are using acids and bases to develop new ways to diagnose and treat diseases. For example, scientists are using acids and bases to develop new drugs that can target and destroy cancer cells.
- Scientists are using acids and bases to study the chemistry of space. Scientists are using acids and bases to develop new catalysts that can be used to split water into hydrogen and oxygen, which can then be used to produce fuel.

Our tour through acid-base chemistry is just an appetizer. Let's continue exploring other important aspects of chemistry in the chapters ahead.

Chapter 4: Organic Chemistry

Introduction

Gear up for a wild ride through the huge and fascinating world of organic chemistry! This branch of chemistry deals with the near-infinite number of carbon compounds crucial for life processes and modern materials. Let's dive in!

What is Organic Chemistry?

Organic chemistry is the study of carbon-based compounds. Carbon is a unique element because it can form four bonds with other atoms, which allows it to create a wide variety of molecules. Organic compounds are found in all living things, and they are also used in a wide variety of products, such as plastics, fuels, and pharmaceuticals.

There are many different classes of organic compounds, including:

- Hydrocarbons: Hydrocarbons are the simplest type of organic compound, and they are made up only of carbon and hydrogen atoms.

- Alcohols: Alcohols contain a hydroxyl group (-OH).
- Aldehydes and ketones: Aldehydes and ketones contain a carbonyl group (C=O).
- Carboxylic acids: Carboxylic acids contain a carboxyl group (-COOH).
- Amines: Amines contain an amino group (-NH2).

Each class of organic compound has its own unique properties. For example, hydrocarbons are typically nonpolar and insoluble in water, while alcohols and carboxylic acids are polar and soluble in water.

Isomerism

Isomerism is a phenomenon in which two or more compounds have the same molecular formula but different structures. There are two main types of isomerism: structural isomerism and stereoisomerism.

Structural isomers have different connectivity between the atoms in the molecule. For example, butane and isobutane are structural isomers of each other.

Stereoisomers have the same connectivity between the atoms in the molecule, but the atoms are arranged differently in space. One type of stereoisomerism is chirality. Chiral molecules are molecules that cannot be superimposed on their mirror image. For example, the amino acid alanine has two chiral isomers, L-alanine and D-alanine.

Organic Reactions

Organic reactions are chemical reactions that involve organic compounds. There are many different types of organic reactions, but some of the most common types include substitution, addition, and elimination reactions.

Substitution reactions are reactions in which one atom or group of atoms in a molecule is replaced by another atom or group of atoms. For example, when methane reacts with chlorine, the hydrogen atom is replaced by a chlorine atom to form chloromethane.

Addition reactions are reactions in which two or more molecules combine to form a single molecule. For example, when ethene reacts with hydrogen, the two molecules combine to form ethane.

Elimination reactions are reactions in which an atom or group of atoms is removed from a molecule. For example, when ethanol is heated with concentrated sulfuric acid, the hydroxyl group is removed to form ethene.

Biomolecules are organic compounds that are essential for life. Some of the most important biomolecules include carbohydrates, lipids, proteins, and nucleic acids.

- Carbohydrates are the body's main source of energy.
- Lipids are used to build cell membranes and store energy.
- Proteins are essential for building and repairing tissues.
- Nucleic acids store and transmit genetic information.

Organic chemistry is a vast and complex field, but it is also one of the most important fields of chemistry. Organic compounds play a vital role in all aspects of life, and organic chemists are working to develop new organic compounds and new ways to synthesize them.

The Magic of Carbon

Carbon's versatility enables millions of compounds. Let's explore some core concepts, by looking closely at the role that Carbon plays:

- Organic compounds primarily contain carbon-hydrogen and carbon-carbon bonds. Carbon backbones provide structure.

- Saturated hydrocarbons like alkanes contain only carbon-carbon single bonds. Unsaturated hydrocarbons have double or triple bonds.
- Alkanes have the general formula C_nH_{2n+2}. Methane CH_4 is the simplest alkane. Properties depend on chain length.
- Alkenes contain a carbon double bond $C=C$. C_2H_4 ethylene has a $C=C$ double bond. Alkenes are more reactive than alkanes.
- Alkynes have a $C\equiv C$ triple bond, like acetylene C_2H_2. Multiple bonds allow more reactivity.
- Cyclic hydrocarbons contain one or more rings. Cyclohexane C_6H_{12} has a 6-carbon ring structure. The fixed shape impacts reactivity.
- Aromatic hydrocarbons contain delocalized pi electrons above and below planar carbon rings. This gives unusual stability.
- Benzene C_6H_6 has a 6-carbon aromatic ring allowing many substitutions while retaining stability. Other aromatics include toluene and naphthalene.
- Petroleum contains many hydrocarbons making it useful as a fuel source and for producing organic compounds. Crude oil refining separates the mixture based on boiling points.
- Functional groups are specific bonding arrangements that define properties. Groups like -OH alcohols readily undergo reactions.
- Alcohols have an -OH group. Phenols have aromatic -OH groups. Ethers have R-O-R groups.
- Aldehydes contain a -CHO group. Ketones have a -CO- group. Carboxylic acids contain -COOH.
- Amines have nitrogen groups -NH2 or -NH-. Amides link nitrogens to carbonyls.
- Esters link oxygen to carbonyl carbon in -COO- groups. Phospholipids have phosphate groups.

Isomerism leads to many possible structures for a molecular formula:

- Structural isomers have the same formula but different bonding structures.

- C3H6 could be propene CH3-CH=CH2 (alkene) or cyclopropane c-C3H6 (cycloalkane).
- Stereoisomers have the same bond connectivity but different 3D orientations.
- Cis and trans isomers differ in positioning of groups across a C=C double bond.
- Optical isomers are mirror images with identical bonding but non-superimposable structures.
- Chirality arises from an asymmetric tetrahedral carbon with 4 different substituents, like your left and right hands. Polarized light rotates differently passing through enantiomers.
- Geometric isomers have restricted rotation about double bonds, fixing groups in space. Cis and trans isomers are geometric isomers.
- Conformational isomers interconvert through rotation about single bonds. Gauche and anti-conformers of butane rotate to interconvert.

Organic reactions rearrange bonds to form new substances. Key reaction classes include:

- Substitution reactions replace an atom or functional group on a hydrocarbon chain with a new substituent:

  CH4 + Cl2 -> CH3Cl + HCl

- Addition reactions join two molecules across a C=C double bond:

  C2H4 + Br2 -> BrCH2CH2Br

- Elimination reactions remove atoms, reforming C=C double bonds:

  2CH3CH2Cl -> 2CH2=CH2 + 2HCl

- Oxidation involves gaining oxygen or losing hydrogen:

$2C_2H_6 + 7O_2 \rightarrow 4CO_2 + 6H_2O$ (combustion oxidizes ethane)

- Reduction gains hydrogen or loses oxygen:

$C_2H_5OH + 2[H] \rightarrow C_2H_6 + H_2O$ (adding H reduces alcohol to alkane)

- Polymerization joins many smaller units into long chain molecules:

$nCH_2=CH_2 \rightarrow -(CH_2-CH_2)n-$ (forming polyethylene polymer)

- Dehydration synthesis removes water to join monomers:

$HOCH_2CH_2OH + HOCH_2CH_2OH \rightarrow$
$HOCH_2CH_2OCH_2CH_2OH$

- Hydrolysis splits bonds using water:

$CH_3COOCH_3 + H_2O \rightarrow CH_3COOH + CH_3OH$

Important types of organic compounds include:

- Hydrocarbons - alkanes, alkenes, alkynes, aromatic rings like benzene are the simplest organics
- Alcohols - contain -OH groups like ethanol CH_3CH_2OH; phenols have aromatic -OH
- Ethers - R-O-R oxygen linkages like methoxymethane CH_3OCH_3
- Aldehydes - have a -CHO group on the carbonyl like formaldehyde CH_2O
- Ketones - carbonyl C=O linked to two carbons like acetone CH_3COCH_3
- Carboxylic acids - contain the -COOH group like acetic acid CH_3COOH

- Esters - have -COO- groups; combine alcohols and acids to form fruity esters
- Amines - have nitrogen groups -NH2 or -NH-; can be primary, secondary or tertiary
- Amides - nitrogen attached to carbonyl carbon like acetamide CH3CONH2

Essential biomolecules for life include:

- Carbohydrates - sugars and polysaccharides used for energy storage (glucose, starch) and structure (cellulose).
- Lipids - fatty acid oils, waxes, phospholipids; provide energy storage, cell membrane structure, signaling molecules.
- Proteins - chains of amino acids that adopt specific 3D shapes to catalyze reactions as enzymes or provide structural elements like muscle, skin, hair.
- Nucleic acids - DNA and RNA chains that store and transmit genetic code, allowing inheritance, protein synthesis, and reproduction.

Some Amazing Facts

Now let's ponder about the wonders of organic chemistry by considering these facts and famous quotes:

- "Organic chemistry is the chemistry of carbon compounds. This is the chemistry that drives life processes." - John E. McMurry
- There are over 10 million known organic compounds to date and likely millions more yet to be discovered. The possibilities are endless!
- Isomerism leads to diverse compounds with the same molecular formula. For example, C3H6 could be propene or cyclopropane.
- Chirality from asymmetric carbon atoms causes molecules to be left-handed or right-handed. DNA is always right-handed.

- Benzene and aromatic compounds have unusual stability due to delocalized pi electrons in the ring. This enables substitution reactions while retaining the ring.
- Organic chemistry arose from the flawed vitalism idea that organic compounds were fundamentally different from inorganic matter. Friedrich Wöhler's synthesis of urea disproved that.
- "The boundary between organic and inorganic chemistry is like a boundary between living and non-living." - Roald Hoffman
- Crude oil and natural gas provide an abundance of hydrocarbon molecules that are separated and turned into fuels and chemicals.
- Important reaction types include addition, elimination, substitution, oxidation-reduction, polymerization and more depending on functional groups.
- Chlorophyll's porphyrin ring structure absorbs visible light for photosynthesis. Small changes to this molecule abolish its light-harvesting ability.
- Amino acids link together through peptide bonds into chains to form proteins with enormously complex three-dimensional folded structures.
- Isoprene polymerizes to form natural rubber, while styrene monomers join to make polystyrene plastics.
- "The high information content of biological molecules like DNA, proteins and carbohydrates is made possible by organic chemistry." - Kurt Wüthrich
- Esters produce pleasant fruity smells, allowing organic chemists to design artificial fragrances and flavors.
- Phospholipids self-assemble into micelles and bilayer membranes in water, enabling cell structure and compartmentalization in biology.
- Every biological transformation - breathing, eating, moving, thinking - runs on the organic chemistry intrinsic to metabolism.
- Organic chemistry has brought about plastics, textiles, fertilizers, pharmaceuticals, dyes, agricultural chemicals and much more.

- "Organic synthesis is like an artistic endeavor seeking to construct intricate molecular masterpieces." - Elias James Corey
- Chirality is crucial in biology and pharmacology since two enantiomers of a drug may have very different effects.
- No other scientific discipline has enabled the design and synthesis of such a vast array of compounds as organic chemistry.
- "Organic chemistry is the chemistry of carbon compounds. Biochemistry is the study of carbon compounds that crawl." - Mike Adams
- "Life is carbon, but carbon is not life. This is the paradox of organic chemistry." - Addison G. Wright
- "If all else fails, immortality can always be assured by spectacular error." - George Bernard Shaw (referring to mistakes in organic synthesis producing novel compounds)
- "Organic synthesis is a highly creative science where chemists think much like architects who assemble structures from molecular building blocks." - Elias James Corey
- "The aromatic sextet of carbon rings in benzene is a unique electronic space, found nowhere else in chemistry, not even on distant stars." - Oliver Sacks

Recent Breakthroughs

- Scientists developed new catalysts that can be used to synthesize organic molecules more efficiently and sustainably. This could lead to new ways to produce pharmaceuticals, plastics, and other chemicals.
- Organic chemists are using new synthetic methods to create complex molecules with unique properties, like solar cells, batteries, and other electronic devices.
- Organic chemists are using new techniques to study the structure and function of biological molecules. This research could lead to new ways to diagnose and treat diseases.

- Organic chemists are developing new ways to produce renewable fuels and other biofuels. This could help to reduce our reliance on fossil fuels.
- Organic chemists are using new methods to study the chemistry of space. For example, scientists have identified organic molecules in meteorites and comets.

Wow, what an incredible diversity of organic compounds and applications! From fuels and pharmaceuticals to the very molecules of life, organic chemistry helps explain our physical world and own existence. Let's continue marveling at the versatility of carbon!

Chapter 5: Nuclear Chemistry

"The nucleus is like an ancient realm, a magic kingdom of protons and neutrons, governed by forces beyond comprehension." - Sam Kean

Introduction

Get ready for a trip into the atomic nucleus, the realm of nuclear chemistry! This mysterious world of subatomic particles, radiation, and immense energy is full of wonder and peril. Let's dive in!

The Nucleus and Radioactivity

Nuclear chemistry is the study of the nucleus of the atom, including its structure, composition, and reactions. The nucleus of an atom is made up of two types of particles: protons and neutrons. Protons have a positive charge, neutrons have no charge, and electrons have a negative charge.

Radioactivity is the spontaneous emission of radiation from the nucleus of an atom. There are three main types of radioactive decay: alpha decay, beta decay, and gamma decay.

- Alpha decay is the emission of an alpha particle, which is made up of two protons and two neutrons.
- Beta decay is the emission of a beta particle, which is an electron or a positron (a positively charged electron).
- Gamma decay is the emission of a gamma ray, which is a high-energy photon.

Nuclear stability is the ability of an atom to resist radioactive decay. Atoms with unstable nuclei will eventually decay to form more stable atoms.

Nuclear Reactions

Nuclear reactions are reactions that involve the nucleus of an atom. Nuclear reactions can be used to create new elements, or to release large amounts of energy.

Transmutation is the process of converting one element into another element. Transmutation can occur through nuclear reactions.

Fission is the process of splitting a heavy nucleus into two lighter nuclei. Fission reactions release a large amount of energy, and they are used in nuclear power plants to generate electricity.

Fusion is the process of combining two light nuclei to form a heavier nucleus. Fusion reactions release even more energy than fission reactions, and they are the process that powers the sun and other stars.

Radioisotopes are isotopes that are radioactive. Radioisotopes have a wide variety of applications in medicine, industry, and research. For example, radioisotopes are used in medical imaging techniques such as PET scans and SPECT scans. Radioisotopes are also used in industrial applications such as food irradiation and smoke detectors.

Some Amazing Nuclear Facts

- Nuclei consist of protons and neutrons held together by the strong nuclear force. Protons are positively charged, neutrons neutral.
- Atomic number is the number of protons, which determines the element identity. Atomic mass is the total protons and neutrons.
- Unstable nuclei undergo radioactive decay, emitting high-energy particles and radiation as they transmute to more stable elements.
- "Radioactivity splits the atom without splitting the nucleus." - James Stokley
- Alpha decay emits an alpha particle, a helium nucleus with 2 protons and 2 neutrons.
  Uranium-238 -> Thorium-234 + He-4
- Beta decay converts a neutron to proton, emitting an electron and neutrino.
- Gamma radiation is a high energy electromagnetic emission causing ionization.
- Half-life is the time for half the radioactive nuclei to decay. Uranium-238 has a half-life of 4.5 billion years!
- "Nuclear reactions convert one element into another, altering protons, neutrons, and energy." - Glenn Seaborg
- Fusion combines light nuclei releasing energy. Hydrogen fuses into helium in stars.
- Fission splits heavy nuclei like uranium or plutonium. This can release huge amounts of energy.
- Transmutation purposefully changes one element into another via nuclear reactions. Alchemists' dream!
- Nuclear weapons like atomic bombs involve uncontrolled chain reactions of nuclear fission to release massive explosions of energy.

- Nuclear reactors control fission reactions in a chain reaction. This heats water to turn turbines and generate electricity.
- "Nuclear fusion promises limitless clean energy if we can harness the power of stars." - David LeBlanc
- Fusion joins hydrogen nuclei into helium at extreme temperatures of millions of degrees.
- The Sun uses proton-proton fusion in its core to release energy.
- Fusion research seeks to replicate solar fusion in containment reactors. Achieving net energy gain has proven very difficult so far.
- Deuterium-tritium fusion is the most feasible reaction to attempt on Earth. High temperatures are needed to overcome the strong nuclear force.
- Inertial confinement and magnetic confinement are leading fusion reactor designs. Exciting progress, but still challenges ahead!
- "Radioisotopes have many innovative applications across science and medicine."
- Radiotracers containing trace radioactive isotopes follow pathways through organisms and environments.
- Carbon-14 dating uses the half-life of carbon-14 to date ancient artifacts and fossils up to 50,000 years old. Archaeologists rejoice!
- Radiation therapy uses targeted radiation from cobalt-60 or linear accelerators to destroy cancer cells.
- Nuclear medicine injects radioactive tracers to image organs and diagnose disease with PET and SPECT scans.
- "The energy produced by the breaking down of the atom is a very poor kind of thing. Anyone who expects a source of power from the transformation of these atoms is talking moonshine." - Ernest Rutherford, 1933
- Nuclear fission produces enormous amounts of energy from small amounts of matter - nearly 1 million times more than chemical reactions per unit mass.
- Radioactive isotopes emitted alpha, beta, and gamma radiation led to the discovery of the neutron by James Chadwick in 1932.

- "Radioactivity is a sign of delicacy, not brute force. It is alchemy, not smashing atoms with a hammer." - Keith McCarthy
- The strong nuclear force overcomes electric repulsion between protons, binding the nucleus together with bond strengths of 100 to 1000 times more than chemical bonds.
- Marie and Pierre Curie pioneered the study of radioactivity and radiation with their work isolating radioactive elements like radium and polonium.
- Nuclear technology arose from the Manhattan Project's work developing atomic weapons during World War 2, including the first nuclear reactors.
- "Nuclear fusion has the potential to provide unlimited clean energy by harnessing the same reactions that power the Sun." - Steven Cowley
- Radioactive isotopes are used in medicine for nuclear imaging techniques like PET scans, and in radiation therapy for cancer treatment.
- The half-life of a radioactive isotope is the time it takes for half the material to decay. Half-lives range from fractions of seconds to billions of years.
- Nuclear spin allows MRI scanning to non-invasively peer inside the human body, providing highly detailed images of soft tissues.
- Radiometric dating uses the predictable decay rates of radioactive isotopes to accurately date ancient objects and geological formations.
- Nuclear power plants provide about 10% of the world's electricity, but also produce hazardous radioactive waste requiring complex long-term disposal.
- Nuclear fusion promises virtually limitless clean energy if the technical challenges of achieving sustained fusion can be solved.
- "The prospect of domination of the nation's scholars by Federal employment is gravely to be regarded." - Albert Einstein on the militarization of nuclear research.
- The critical mass for a self-sustaining fission chain reaction depends strongly on the type of fissile isotope and conditions like material density and shape.

- Nuclear transmutation artificially induced nuclear reactions to convert one element into another, such as producing plutonium for nuclear weapons.
- Radiotracers incorporating radioactive isotopes allow tracking of chemicals through organisms or industrial processes.
- Nuclear forensics uses radiation signatures to identify origins and smuggling of nuclear materials.
- Nuclear technology will continue playing an important yet controversial role in energy, medicine, industry, transportation, and defense.

Nuclear chemistry is a fascinating and important field of science. Nuclear chemistry has led to the development of many new technologies, and it continues to be an active area of research.

Recent Breakthroughs

- Recently, in 2020, scientists got the first glimpse of elusive xenon tetroxide, an exotic compound useful for next-gen nukes.
- 2021 unleashed element 117, Tennessine, blasting its way onto the periodic table.
- 2022 saw nuclear batteries powered by diamonds, while physicists cooked up the most precise atomic clock ever using strontium atoms. 2023 took the radioactive cake with compact fusion reactors heating plasma to 180 million degrees and lighting the way to limitless clean energy.

We've only scratched the surface of the mysteries of the atomic nucleus! Let's continue exploring the incredible world of nuclear chemistry.

Chapter 6: Thermodynamics and Kinetics

Introduction

Thermodynamics and kinetics are two key branches of physical chemistry that help explain and predict how chemical systems behave. Thermodynamics deals with energy changes during chemical reactions and equilibria. Kinetics focuses on reaction rates and mechanisms. This important duo provides insights into properties like temperature, pressure, heat transfer, equilibrium constants, catalysts, and more. Let's explore some of the core concepts of thermodynamics and kinetics!

The Laws of Thermodynamics

Thermodynamics relates heat and other forms of energy transfer to chemical systems. The laws of thermodynamics describe fundamental constraints:

- The zeroth law defines temperature and thermal equilibrium. If Object A is in equilibrium with B, and B is in equilibrium with C, then A is also in equilibrium with C. This allows temperature measurements.

- The first law states that energy is conserved in a closed system. The change in internal energy equals heat added minus work done by the system.
- The second law says that entropy or disorder increases in the universe. Spontaneous processes increase entropy.
- The third law deals with the concept of absolute zero.

Law	What It Says
Zeroth Law	If two things are the same temperature as a third thing, they're the same temperature as each other.
First Law (Conservation of Energy)	Energy can't just appear or disappear; it can only change forms. It's like a book that can be read but not created or destroyed.
Second Law (Entropy)	Things naturally tend to get messier over time. Energy spreads out and becomes less useful. Think of a neat room getting cluttered.
Third Law (Absolute Zero)	You can't reach absolute cold (colder than freezing) in any number of steps. It's like trying to cool something down forever, but you'll never get there.

Table. The Laws of Thermodynamics

Beyond these fundamental laws, thermodynamics provides extremely useful parameters for describing energy flows:

- Enthalpy H relates heat content at constant pressure. The change in enthalpy ΔH indicates how much heat is produced or absorbed in a reaction.
- Entropy S measures molecular randomness and disorder. Reactions favor increasing entropy.
- Gibbs free energy G combines enthalpy and entropy to predict spontaneity. Reactions proceed spontaneously if they decrease G.
- Equilibrium constants derived from thermodynamics reveal the extent of reaction and formation of products. Le

Chatelier's principle describes how systems shift to relieve stress and reach new equilibria.

While thermodynamics describes energy states, reaction kinetics focuses on reaction rates:

- Reaction rates depend on temperature, pressure, concentrations, and other factors per the collision theory. More collisions means faster reactions.
- Rate laws describe how the rate depends on reactant concentrations. Rate = $k[A]x[B]y$ for reaction orders x and y.
- The Arrhenius equation gives the temperature dependence of the rate constant k.
- Activation energy Ea is the energy barrier to overcome for a reaction to proceed. Catalysts lower Ea to speed up reactions.
- Reaction mechanisms outline the stepwise sequence of bond making and breaking underlying the overall reaction. Fast steps limit the rate.
- Theories like transition state theory provide further molecular insights into kinetics.

Some Amazing Facts

Let's marvel at some thermodynamic and kinetic phenomena and consider some famous quotes:

- "Thermodynamics is the only physical theory of universal content which I am convinced will never be overthrown." - Albert Einstein
- The zeroth law of thermodynamics allows the definition of temperature and thermal equilibrium. Objects in contact will reach the same temperature.
- Entropy is a measure of molecular disorder and randomness. The second law of thermodynamics states that entropy in an isolated system always increases.

- "Thermodynamics governs the unspoken laws of existence. Its principles are written into the cosmos." - Michio Kaku
- Enthalpy (H) describes the total heat content of a system at constant pressure. The change in enthalpy (ΔH) indicates how much heat is evolved or absorbed during a reaction.
- Gibbs free energy (G) accounts for both enthalpy and entropy driving forces. Reactions proceed spontaneously if they result in a decrease in G.
- Equilibrium constants calculated from thermodynamic data reveal the extent to which a reaction proceeds towards products vs reactants at equilibrium.
- Transition state theory aims to explain reaction rates based on the thermodynamic properties of the transition state complex along the reaction coordinate.
- The Arrhenius equation gives the quantitative dependence of the rate constant k on the activation energy Ea and temperature T.
- "Kinetics teaches us that chemical reactions need time to unfold their mysteries." - Istvan Hargittai
- Reaction mechanisms outline the sequence of individual steps by which reactants are converted into products during a chemical reaction.
- Catalysts increase reaction rates by lowering the activation energy barrier without being consumed in the process.
- Reaction rates depend on factors like temperature, pressure, and concentration that affect molecular collisions and orientation.
- Rate laws describe the mathematical relationship between reaction rate and species concentrations.
- The carbonate bicarbonate buffering system in the ocean helps regulate Earth's climate by absorbing carbon dioxide.
- Life itself likely emerged through complex chemical kinetics on the early Earth under ideal conditions.
- "An organism's astonishing ability to convert sugar into energy is a triumphal achievement in kinetics." - Fritz Haber
- Chemical clocks like the Briggs-Rauscher oscillating reaction demonstrate how kinetics can produce surprising cyclical color changes over time.

- Chain reactions involve self-propagating kinetic steps, like in nuclear fission reactions.
- Chemical thermodynamics and kinetics continue providing deep insights into nature at scales from atoms to galaxies.
- "Thermodynamics explains why you can't win, kinetics is how fast you lose." - Ken Dill
- Burning gasoline in an engine is an irreversible process increasing entropy. The released heat can do useful work.
- Evaporation cools objects by the enthalpy of vaporization. Sweating uses evaporation to cool us down!
- Le Chatelier's principle explains how added heat or pressure will shift equilibria. High pressure favors smaller gas volumes.
- Enzymes speed up biological reactions by lowering activation energy. Their mechanisms are complex and finely tuned by evolution.
- Reaction rates roughly double with each 10°C rise in temperature. Thermite reactions get incredibly fast!
- Catalytic converters use platinum and rhodium to minimize car emission pollution through catalysis.

Recent Breakthroughs

- In 2020, chemists crushed diamonds to observe metastable polymeric carbon forms, exploring exotic high-pressure chemistry.
- 2021 saw microscale thermometers map energy flows in bacteria with unprecedented resolution.
- In 2022 thermoelectric materials made breakthroughs, providing new sustainable heat harvesting options.
- In 2023, AI and machine-learning algorithms were predicting intricate chemical mechanisms and blazing new trails in computational chemistry.

The science of energy, order, chaos and speed shows no signs of slowing down! The intricacies of thermodynamics and kinetics illuminate everything from meteorology to metabolism. Let's continue exploring!

Chapter 7: Electrochemistry and Redox Reactions

Introduction

Electrochemistry involves the interplay of electricity and chemical reactions. Oxidation-reduction (redox) reactions are central to electrochemistry. When redox reactions are separated spatially, electron transfer occurs through an external circuit producing electrical energy. This allows batteries, fuel cells, electrolysis, and more technologies. Let's explore some core electrochemistry concepts!

Oxidation and reduction are complementary processes involving electron transfer. Oxidation means losing electrons, reduction means gaining electrons. Redox reactions always involve both oxidation and reduction occurring simultaneously.

- Oxidation states or oxidation numbers track the hypothetical charge on individual atoms if all bonds were fully ionic. Useful for balancing redox reactions.

- Oxidation increases the oxidation state. Reduction decreases the oxidation state. Atoms accepting electrons are reduced.
- Strong reducing agents easily donate electrons. Strong oxidizing agents readily accept electrons.
- Common oxidizing agents: oxygen, chlorine, permanganate, dichromate.
- Common reducing agents: lithium, sodium, hydrogen gas, carbon monoxide.

Electrochemical cells harness redox reactions to produce electricity:

- In galvanic or voltaic cells, spontaneous redox reactions drive electron flow, generating power.
- At the anode, oxidation occurs releasing electrons that flow through the external circuit.
- At the cathode, reduction occurs absorbing electrons coming from the external circuit.
- Salt bridges maintain charge balance between the two half cells.
- Cell potential relates to the electrode potentials of the two half reactions. Higher potential differences produce more voltage.
- The Nernst equation gives the cell potential at non-standard conditions based on concentration, pressure, and temperature.
- Batteries contain multiple galvanic cells connected in series to provide higher voltages. Alkaline, lead-acid, and lithium ion batteries are common.
- Fuel cells oxidize fuels like hydrogen gas electrochemically to generate power. The oxygen reduction half reaction occurs at the cathode.

Electrolytic cells use electricity to drive non-spontaneous redox reactions:

- Electroplating applies a voltage to reduce metal cations, depositing a thin metal coating on a surface.
- Electrolysis of water splits water into hydrogen and oxygen gas. $2H2O + electricity \rightarrow 2H2 + O2$
- Hall-Heroult process electrolyzes aluminum oxide to highly reactive aluminum metal.

Some Amazing Facts

- "All of the batteries in the world connected together could only power the world for about 15 minutes."
- The voltaic pile, invented by Alessandro Volta in 1800, was the first electrical battery that produced a reliable flow of electric current.
- The standard hydrogen electrode serves as the reference point for all other electrode potentials, defined as 0.00 V at standard conditions.
- "Every second breath you take comes from photosynthesis and the electrochemical splitting of water by plants." - John Bockris
- Electroplating allows coatings of metal as thin as one micron - about 1/25 the thickness of a human hair!
- The Hall-Heroult process uses huge amounts of electricity to produce aluminum metal from bauxite ore through electrolysis.
- Batteries rely on electrochemistry, with reduction and oxidation reactions separated between the cathode and anode.
- Lithium ion batteries power most portable electronics today. They have high energy density but risk flammability and degradation over time.
- "Electrochemistry serves as the foundation for many emerging energy technologies." - Kevin Huang

- The standard hydrogen electrode provides a universal reference for measuring other electrode potentials. It's assigned 0.00 volts at standard conditions by definition.
- Commercial metal refining and extraction processes harness electrochemical reactions for efficiency and speed.
- Chlorine gas was first produced through the electrolysis of brine by Humphry Davy in the early 1800s.
- Electroplating allows deposition of a thin layer of metal like chromium or nickel to provide corrosion protection or decoration.
- The Nernst equation describes how cell potential varies with temperature, pressure and concentration.
- "Batteries convert chemical energy directly into electrical energy with high efficiency." - Vincent Sprenkle
- Electrochemical systems like batteries and fuel cells will play a key role in transitioning from fossil fuels to renewable energy.
- Alkaline batteries rely on the reduction of manganese dioxide and zinc reactions. Their voltage output declines as they are discharged.
- Early batteries were known as "voltaic piles" since they contained stacked plates or cells that produced electricity.
- The lead-acid battery still used in most cars today was invented by French physicist Gaston Plante in 1859.
- Electrochemistry enables widespread technologies from batteries and sensors to water treatment and metal production.

Recent Breakthroughs

Electrochemistry has many important applications! The pace of discoveries shows no sign of stopping down. Just like the Energizer Bunny!

- Electrochemical energy storage devices are getting better and more efficient. In 2022, a team of scientists developed a new type of lithium-ion battery that can charge in just 5 minutes and last for over 20 years. This could revolutionize the electric vehicle industry and make renewable energy more practical.

- Electrochemistry is being used to develop new ways to produce clean energy. In 2021, researchers at the University of California, Berkeley, developed a new type of solar cell that uses electrochemistry to split water molecules into hydrogen and oxygen. The hydrogen can then be used to store energy or produce fuel.

- Electrochemistry is also being used to develop new ways to remediate pollution. In 2020, a team of scientists at the University of Toronto developed a new electrochemical process that can remove PFAS, a class of persistent organic pollutants, from water. This could help to clean up contaminated groundwater and drinking water supplies.

- Electrochemistry is being used to develop new medical treatments. In 2023, a team of researchers at Harvard University developed a new electrochemical device that can be implanted in the body to deliver insulin to patients with diabetes. This device could help to improve the lives of millions of people with diabetes.

- Electrochemistry is being used to develop new materials with unique properties. In 2022, a team of scientists at the University of Manchester developed a new type of material that can be used to create batteries that are both more efficient and more durable than existing batteries. This material could also be used to develop new types of sensors and other electronic devices.

Chapter 8: Spectroscopy and Analytical Methods

Introduction

Spectroscopy involves studying the interaction between light and matter. When atoms or molecules absorb or emit light at specific wavelengths, their spectra can provide information about composition and chemical structure. Various spectroscopic techniques are invaluable for analyzing materials in diverse fields from astrophysics to forensic science. Let's explore some key concepts in spectroscopy!

Light acts as both a particle and wave, with properties described by quantum physics. Different wavelengths of light relate to different quanta of energy per Planck's equation.

- The electromagnetic spectrum categorizes types of light based on wavelength and frequency, spanning cosmic rays, gamma rays, X-rays, ultraviolet, visible, infrared, microwaves, and radio waves.

- When atoms absorb light, electrons get promoted to higher energy levels. Emission occurs when electrons relax back down, emitting photons at characteristic wavelengths.
- Atomic spectra display bright emission or absorption lines corresponding to the quantized electronic transitions in atoms. Elemental fingerprints!
- Molecular spectra also show absorption or emission bands relating to vibrations and rotations of chemical bonds. Bonding structure affects the spectrum.

Spectroscopic techniques leverage these quantum interactions for analysis:

- Atomic absorption spectroscopy measures light absorption to determine analyte concentration.
- Fluorescence and phosphorescence spectroscopy examine molecular relaxation for structure insights.
- Vibrational spectroscopy including infrared and Raman spectroscopy reveals bond motions.
- Nuclear magnetic resonance (NMR) spectroscopy utilizes EM pulses and spin states to identify molecular structure.
- Mass spectrometry weighs molecular and fragment ion masses. The mass spectrum fingerprints chemical identity.
- X-ray diffraction and spectroscopy probe inner shell electrons to determine electron configurations and bond distances.

Other common spectroscopic methods:

- UV-Vis spectrophotometry measures absorption of UV and visible light.
- Atomic emission spectroscopy detects emitted wavelengths.
- Photoelectron spectroscopy ejects electrons to measure binding energies.

Advanced spectroscopic techniques shed light on the smallest scale structures and improve our understanding across scientific fields. Let's get spectral!

Some Amazing Facts

- "Spectroscopy allows us to use light to uncover the secrets of matter." - Brian Rohrig
- Hydrogen produces the simplest atomic emission spectrum with four visible lines, as its single electron transitions between energy levels.
- The Lyman series of ultraviolet absorption lines in hydrogen's spectrum was discovered in 1906 by Theodore Lyman.
- Each element has a unique atomic emission spectrum like a fingerprint used to identify it in samples of unknown composition.
- X-ray diffraction techniques exploit the wave properties of electrons to determine crystal structures and molecular conformations.
- "Infrared spectroscopy probes the invisible world of molecular vibrations and rotations." - Peter Griffiths
- NMR spectroscopy utilizes the magnetic properties of atomic nuclei to study molecular structure and composition.
- Edwin Hubble used redshifts in the spectra of other galaxies to show the universe is expanding.
- Mass spectrometry separates and analyzes ions based on their mass-to-charge ratio, generating a plot of ion abundance vs mass.
- Chromatography separates mixtures by differential partitioning between a mobile and stationary phase.
- Atomic force microscopy uses an extremely fine tip to scan surfaces and measure tiny forces, rendering nanoscale topographical images.
- "Spectroscopy arose from the interaction of light, quantum theory, and human curiosity." - Stuart M. Lindsay

- Absorption and emission spectra were pivotal evidence for the quantum mechanical model of atoms with quantized energy levels.
- Spectrophotometry measures how much light is absorbed by a sample at different wavelengths to quantify composition.
- Flame emission spectroscopy sprays a sample into a flame to excite atomic emissions indicating metal ions present.
- Raman spectroscopy analyzes inelastic scattering of monochromatic light, such as from a laser source, revealing vibrational modes.
- "There is no spectrum more beautiful than that which radiates from a mind illuminated by spectroscopic knowledge." - Kurt Wüthrich
- Photoelectron spectroscopy measures the binding energies of electrons by ejecting them with photons and analyzing their kinetic energy.
- Mössbauer spectroscopy examines nuclear transitions to study chemical environments and bonding.
- Spectroscopic advances have enabled astronomers to analyze the composition of stars light years away!

Recent Breakthroughs

- Scientists have developed new spectroscopic techniques that can detect and identify molecules at the single-molecule level. This could revolutionize our understanding of chemical reactions and biological processes.
- Spectroscopy is being used to develop new ways to diagnose and treat diseases. For example, scientists are using spectroscopy to develop new blood tests that can detect cancer cells and other biomarkers of disease.
- Spectroscopy is also being used to develop new ways to monitor the environment. For example, scientists are using spectroscopy to develop new sensors that can detect air and water pollution.

- Spectroscopy is being used to develop new materials with unique properties, like new materials for solar cells and batteries.
- Scientists are using spectroscopy to study the atmospheres of exoplanets and to identify organic molecules in meteorites.

That concludes our journey into the vast and illuminating field of chemistry! You are now a certified CHEMISTRY NERD!

Now, stay tuned for a FREE excerpt from one of our bestselling books, AI for Smart Pre-Teens and Teens Ages 10-19. I hope you will consider adding it to your library.

FREE SAMPLE from this book coming up NEXT!

Get on the leading edge of the AI Revolution!
For pre-teens, teens and older

SCAN ME

- AI demystified: learn the fundamentals behind AI, including machine learning, neural networks, and deep learning
- Get inspired by industry leaders like Sam Altman, Elon Musk and Satya Nadella
- Help drive the responsible and beneficial use of AI
- Perfect travel companion or gift

Follow Dr. Leo Lexicon on Twitter/X
X @LeoLexicon

LEXICON LABS

FREE SAMPLE Chapter (Ch.5 From the Mouths of Experts)

Artificial intelligence is shaping the future, but it didn't happen overnight. Over the years, the development of AI has benefited from the genius of many people. We will hear from some of the top innovators who helped shape AI in this chapter. Prominent individuals from a range of businesses have different viewpoints on the possible implications and difficulties of artificial intelligence (AI), a topic that is fast growing. While this list is not exhaustive, it is always important to identify and understand the viewpoints of key people in any domain. Here is a more detailed look at their opinions:

Sam Altman, OpenAI's CEO

Sam Altman is a well-known name in the AI field, and he has a very upbeat outlook on how AI could change the world. According to Altman, AI has the potential to outperform even the revolutionary effects of early breakthroughs like electricity and fire. He imagines a time when AI will be used as a potent weapon against serious global problems like sickness and poverty. In Altman's opinion, the proper use of AI may change industries, boost productivity, and enhance people's quality of life all across the world.

Altman does acknowledge that there are considerable hazards associated with such enormous potential. He is fully cognizant of the potential ethical, social, and economic difficulties that AI may present. Altman highlights the significance of ensuring that the advantages of AI are available to everyone, regardless of socioeconomic class or geographic location, as AI develops. In order to prevent escalating already existing inequities, he supports regulations that encourage a just and equitable sharing of AI's benefits.

Elon Musk, Founder of Tesla

Elon Musk has a long connection with AI, and he was in fact one of OpenAI's co-founders, but later dropped out of the company, long before it released ChatGPT. Elon's views suggest that he approaches AI with more caution. He acknowledges that AI has the potential to improve a variety of industries, but he is also gravely concerned about the risks that could arise from unrestrained AI research. He has even compared AI to nuclear weapons in an effort to warn that, if not properly governed, it could endanger humanity.

The lack of regulatory control in AI research is Musk's main worry. He thinks that in the absence of appropriate rules and regulations, AI systems might be used without sufficient safety precautions, which could have unforeseen repercussions. Musk is a supporter of strict governance and ethical frameworks that put safety and human values first in order to ensure that AI is developed ethically. In order to ensure that AI research and application are carried out with the highest caution and responsibility, he argues for the creation of institutions and policies.

Musk recently announced the creation of a new company called X AI devoted to responsibly advancing artificial intelligence. The company will focus on creating transparent AI systems that can articulate their decision-making processes in a way humans can understand. Musk said X AI will allow AI developers "to see how the AI thinks and why it makes certain decisions." He hopes X AI will set a new gold standard for ethical, thoughtful AI design that other companies will follow, leading to AI that truly augments human abilities. Given Musk's profile, X AI is sure to quickly become an influential player in the high-stakes world of AI safety research.

Yoshua Bengio, Deep Learning Pioneer

Deep learning pioneer Yoshua Bengio provides insight into the state of AI today and its limits. He calls AI "narrow and brittle," emphasizing how most AI systems do well at performing narrowly defined tasks but fall short when it comes to generalizing knowledge and responding to novel circumstances.

Artificial general intelligence (AGI), which refers to AI systems with the capacity to reason and comprehend the world at a level equivalent to human intelligence, is a goal strongly supported by Bengio. He thinks that in order to attain AGI, it is crucial to learn more about how the human brain works and how neural networks and biological processes contribute to intelligence.

Bengio believes that cognitive and neuroscience research is essential for directing the creation of AI systems that can learn, generalize, and adapt just like people. In order to create more adaptable and flexible AI systems, he promotes multidisciplinary collaborations between specialists in AI and other scientific domains.

"The Godfather of deep learning" Geoffrey Hinton

One of the pioneers of deep learning, Geoffrey Hinton is internationally recognized for his contributions to the development of AI. Hinton's viewpoint focuses on the necessity of updating AI datasets and algorithms to produce more intelligent machines. He berates modern AI models for not having a thorough comprehension of reality. Although AI has made great progress in a number of specific tasks, Hinton contends that these systems frequently lack the reasoning and knowledge generalization skills necessary for attaining real intelligence. He urges the use of fresh methods in AI research that could help us comprehend the fundamentals of human intelligence better.

Hinton imagines a time where AI helpers can relate to people as dependable friends and have empathy for them. He thinks that for AI to succeed, it must advance beyond simple pattern recognition and gain comprehension of context, emotions, and intentions. In Hinton's perspective, AI can be a real collaborator who can connect with people on a deeper level rather than just a tool for particular jobs.

Fei-Fei Li, Co-director of Stanford's AI Lab

Fei-Fei Li, a well-known researcher and lecturer in the field of AI, believes that AI has enormous potential to enhance human potential and improve life. She is aware of how AI can revolutionize many industries, such as healthcare, education, and environmental sustainability. AI is a useful tool for tackling complex problems and improving scientific study because of its capacity to process enormous volumes of data and spot patterns.

However, Li stresses that it is crucial that AI is developed in an ethical and responsible manner. She is an advocate for greater diversity in the AI community since it can lessen the negative biases present in the technology. To prevent sustaining societal disparities, it is essential to ensure inclusivity and justice in AI applications.

Li is a strong supporter of the viewpoint that AI should be created with an emphasis on enhancing rather than replacing human capabilities. She thinks AI systems ought to be created so they may coexist peacefully with people, boosting their skills and enabling them to make wiser judgments. AI may be used as a potent tool to address global concerns and enhance the quality of life for all people by keeping humans in the loop and prioritizing human-AI collaboration.

Andrew Ng, Google Brain Co-Founder

The well-known AI researcher and entrepreneur Andrew Ng is certain that AI has the power to fundamentally alter the healthcare industry. He views artificial intelligence as a useful tool that can help with earlier and more precise disease diagnosis, perhaps saving lives. AI can help doctors make better judgments and improve patient outcomes by analyzing massive volumes of medical data and looking for trends.

Ng warns against overestimating AI's potential, though. While AI has demonstrated tremendous accomplishments in certain jobs, it still lacks the human ability for common sense and general intelligence. Due to this constraint, AI might be excellent in specific fields yet struggle to comprehend complicated real-world situations that call for human-level comprehension and reasoning.

Demis Hassabis, DeepMind's CEO

Demis was the leader of the team that developed AlphaGO, the AI that famously defeated the GO world champion, Lee Sedol. Demis imagines a time when AI and people work in unison to solve the world's problems. He understands that AI is particularly good at digesting large volumes of data and making judgments based on that data. Humans, on the other hand, contribute special abilities like compassion, imagination, and intuition to the table. Hassabis thinks we can better tackle complicated issues if we combine the analytical strength of AI with human qualities.

Under Hassabis' direction, DeepMind has been aggressively examining how AI could help professionals in a range of industries, including healthcare and scientific research. Hassabis's idea of AI-human synergy is demonstrated through the company's partnerships with medical organizations to develop AI systems that assist in disease diagnosis and drug research.

Cynthia Breazeal, Creator of Jibo

A pioneer in the field of human-robot interaction, Cynthia Breazeal focuses on developing social robots that can comprehend and relate to people emotionally. She is adamant that in order for society to accept AI, it must be more than just a machine that performs tasks; it must exhibit human-like traits that enable sincere emotional interactions.

The goal of Breazeal is to create AI-enabled robots that can perceive and react to human emotions, allowing them to develop deep connections with their users. Such sympathetic AI can be used in a variety of industries, including healthcare, where social robots can help patients emotionally, and education, where they can effectively engage and motivate students.

Eliezer Yudkowsky, Machine Intelligence Research Institute

Eliezer Yudkowsky is a fervent supporter of AI ethics and safety. He issues a warning regarding the dangers that could arise from the creation of complex AI systems. A lack of appropriate prudence and ethical standards could have unforeseen repercussions or even threaten humanity's existence.

To ensure that AI systems behave ethically, Yudkowsky highlights the significance of matching AI's objectives with human ideals. Yudkowsky thinks we can prevent situations in which AI can unintentionally injure people or behave against their best interests by giving safety a higher priority in AI research and development.

Ilya Sutskever, Co-Founder of OpenAI

Ilya Sutskever is one of the co-founders of OpenAI. He asserts that artificial intelligence (AI) may overtake human intelligence within a few decades. One of humanity's biggest concerns may be how to transition to digital superintelligences.
Sutskever underlines the significance of incorporating human values into the design of new AI systems. As AI develops into superintelligence, it is essential to make sure that these potent agents act in accordance with human interests and prevent any unanticipated negative outcomes.

Satya Nadella, CEO of Microsoft

In order to avoid relying on "black box" solutions, Satya Nadella urges businesses to embrace AI as a core skill. Satya Nadella is adamant that AI will change every industry. In the future, he sees AI influencing many facets of corporate operations, from boosting productivity to tailoring customer experiences.

Nadella is dedicated to democratizing AI so that it is available to people and organizations of all sizes. He sees a world where AI platforms and tools are accessible to everyone and are easy to use, allowing more people to use AI to solve issues and spur creativity.

Yann LeCun, Facebook AI Research

Yann LeCun Leading expert in AI Yann LeCun is a proponent of AI systems that enhance and supplement human abilities. LeCun contends that creativity and common sense, which present AI systems lack, are essential components of intelligence.

LeCun advises concentrating on creating AI systems that can collaborate with people in a positive way in order to produce AI that can actually improve human intelligence. In order to provide AI systems the ability to learn more independently and comprehend complicated real-world settings better, research is being done in areas like reinforcement learning and unsupervised learning. LeCun sees a time when artificial intelligence (AI) bridges the gap between human and machine intelligence, enhancing human capabilities and assisting in the solution of humanity's most pressing problems.

Mo Gawdat, Former Chief Business Officer, Google X

Mo Gawdat, also the author of a popular book, Solve for Happy, has an optimistic yet cautious perspective on artificial intelligence. He believes AI has immense potential to automate jobs and tasks, freeing up human creativity and time for higher pursuits. However, Gawdat also recognizes the risks of superintelligent systems and the need to align advanced AI with human values and ethics. He advocates for policies to smooth the transition and distribute the benefits, along with designing AI thoughtfully with emotional intelligence and compassion in mind. Overall, Gawdat argues we can harness the power of AI to improve lives if we steward it carefully, create wise governance, and democratizing access to shape an ethical, benevolent AI that enhances human potential while mitigating the existential risks.

A Final Word

I truly appreciate your participation in this unique journey through the mysteries of chemistry. Please help me spread the word by:

- Leaving a 5-star review at the store where you purchased this book.
- Telling your siblings, classmates, friends and relatives about this book
- Recommending this book to your teacher, coach or educator, and
- Sharing your thoughts on social media

I hope you also liked the free sample from the book, AI for Smart Pre-Teens and Teens Ages 10-19. Do check out our other exciting titles - some of them are highlighted in the pages that follow.

Finally, don't forget to sign up to our newsletter and download your FREE poster print! Go to https://mindzen.squarespace.com/ and sign up today!

I wish you lots of good luck and new adventures!

Dr. Leo Lexicon

Explore the lives of great innovators, Scientists, Leaders, Artists and Explorers...Stay tuned for additional titles coming soon!

Learn the basics of Coding and program in Python.
No prior knowledge required!

Learn all about starting and growing a business <u>as a teenager</u>

Explore the Future of Quantum Computing

FUN BOOKS FOR TRIVIA NIGHT

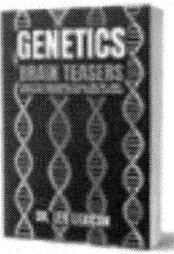

TEST YOUR INNER NERD!

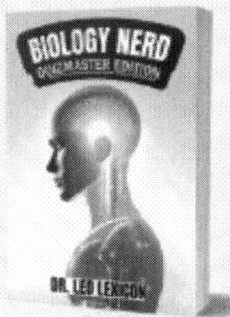

COLORING BOOKS

Discover More Bestselling Titles
from Lexicon Labs!

SCAN ME

Printed in Great Britain
by Amazon

54396905R00044